EXPLORING MECHANICS

The MEW Group

HODDER AND STOUGHTON
LONDON   SYDNEY   AUCKLAND   TORONTO

Authors

Charles Ardagh	Chesham High School
John Berry	Polytechnic South West, Plymouth
Alan Davies	Hatfield Polytechnic
Roger Fentem	Devonport High School for Girls, Plymouth
Ted Graham	Polytechnic South West, Plymouth
Peter Hudson	Teesside Polytechnic
Janet Jagger	Trinity and All Saints College, Leeds
Brian Lowe	Polytechnic of Wales
Stewart Townend	Liverpool Polytechnic
John Walton	Monks Walk School, Welwyn Garden City
Roger Whitworth	Droitwich High School
Julian Williams	Manchester University

British Library Cataloguing in Publication Data

Exploring mechanics
   Problems and investigations
   1. Mechanics
   531

ISBN 0 340 49933 8

First published 1989

© 1989 The MEW Group

All rights reserved. No part of this publication may be reproduced or transmitted in any form or by any means, electronic or mechanical, including photocopy, recording, or any information storage and retrieval system, without permission in writing from the publisher or under licence from the Copyright Licensing Agency Limited. Further details of such licences (for reprographic reproduction) may be obtained from the Copyright Licensing Agency Limited, of 33–34 Alfred Place, London WC1E 7DP.

Typeset by Gecko Limited, Bicester, Oxon.
Printed in Great Britain for the educational publishing division of Hodder and Stoughton Ltd, Mill Road, Dunton Green, Sevenoaks, Kent by Thomson Litho, East Kilbride.

The following companies and institutions have given permission to reproduce photographs in this book: Allsport UK Ltd (35, right and left, 36), Autoglass (34), British Road Federation (25).

Contents

Introduction		iv
Part 1:	What happens if . . . ?	1
Part 2:	Short Investigations	
1	The two-second rule	24
2	An experiment for reaction times	24
3	Motorway traffic jams	25
4	Road humps	26
5	The amber gambler	26
6	Design a road-block	27
7	Bird-brained mechanics	28
8	Artificial gravity	28
9	Down like a lead balloon	29
10	Bicycle gearing	29
Part 3:	Long Investigations	
11	The sliding ruler	32
12	Tumble driers	32
13	Design an accelerometer	33
14	Water chute mechanics	33
15	Convoys	34
16	Smashing windscreens	34
17	Touchdown	35
18	Catch it!	35
19	Parachutist's dilemma	35
20	Sprint starts	36

Introduction

The many applications of mechanics make it an important topic of interest to applied mathematicians, scientists and engineers. For example, the vibrations of structures and the need to determine the resonant frequencies are of fundamental importance to the industrial vibration modeller. The design of aircraft and cars requires a good understanding of momentum and energy. The forces acting on bodies travelling in circular paths have applications to spin driers and orbiting satellites.

Think of an object moving in a complicated way; it might be a cricket ball spinning through the air; a swimmer sliding down a water chute; a car approaching a set of traffic lights. Why does the object move in the way it does? How does the design of the water chute affect the swimmer? Does the car have to stop at the traffic lights when they turn to amber?

Questions like these are often very important to answer and the laws of mechanics are the starting point for investigating such problems. When you learn the principles of mechanics it is important, as a starting point, to clear away the clutter of the real world. You should consider particles and rigid bodies behaving in rather idealised situations. With the knowledge gained from these situations and with a sound understanding of the theory you can apply laws of mechanics to model more realistic situations and successfully predict the outcome.

The set of problems in this book are designed with two aims:
(i) to provide a set of problems to test your understanding of basic concepts, and help you to avoid some of the most common errors made in mechanics,
(ii) to provide some realistic problems to investigate which will illustrate the power of the laws of mechanics you are learning.

In tackling some of these problems it is inevitable you will become bogged down. Don't worry too much. This is an important part of problem solving. Perhaps you can simplify your approach, as after all, a simple model is often better than no model at all. There is tremendous satisfaction in achieving a solution to realistic problems.

We hope that you enjoy dipping into each part of this book, that you gain a better understanding of the concepts of mechanics and that consequently you see a greater relevance to the study of mechanics.

John Berry

PART 1

1

What happens if . . . ?

Problem 1

What happens to you if you step off a moving bus on to the pavement? Give reasons for your answer.

Problem 2

The diagram below shows the positions, at one second time intervals, of two table tennis balls rolling along parallel tracks.
Do the balls ever have the same speed?

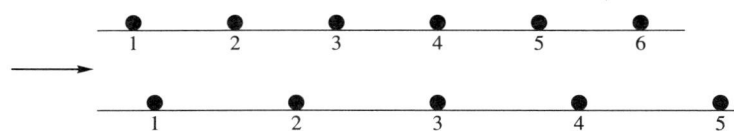

Problem 3

A car is travelling at constant speed along a winding road. Is the acceleration zero, constant or variable?

Problem 4

A ball is dropped out of a window of a car which is travelling along a straight road. Where does the ball land if the car is
(a) moving at constant speed,
(b) slowing down?

Problem 5

While travelling in the back of a pick-up truck, being driven at constant speed, a boy throws a ball straight up into the air.
Can he catch the ball if the car is travelling
(a) along a straight road,
(b) round a bend?

Problem 6

A piece of string is tied to a pedal of a bicycle as shown in the photograph. Which way does the bicycle move? Explain your answer.

Problem 7

A spaceship in deep space, where no forces are acting on it, drifts sideways from X to Y. When it reaches Y it fires its engines for a short time, turning them off when it reaches Z.
Draw a possible path that the spaceship could follow from Y to Z and then beyond Z.

Problem 8

A channel in the shape of a spiral lies on a horizontal table. A small ball is projected along the spiral from the centre. When the ball leaves the spiral at point B, what is its direction of motion? Sketch the path of the ball from point A on the channel through point B and its path on the table.

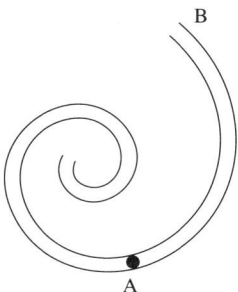

Problem 9

A spaceship in deep space has two engines on full thrust. What happens if engine A runs out of fuel?

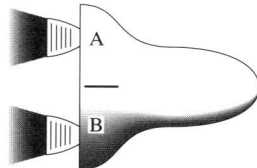

Problem 10

An ice puck slides over the ice with constant speed. What happens if it is struck with an ice-hockey stick in a direction at right angles to its direction of motion? Show on a diagram the motion of the puck before and after it is hit.

Problem 11

A simple experiment to demonstrate the reluctance of an object to move is shown in the photograph. It shows a coin resting on a card.
What happens if the card is flicked sharply?

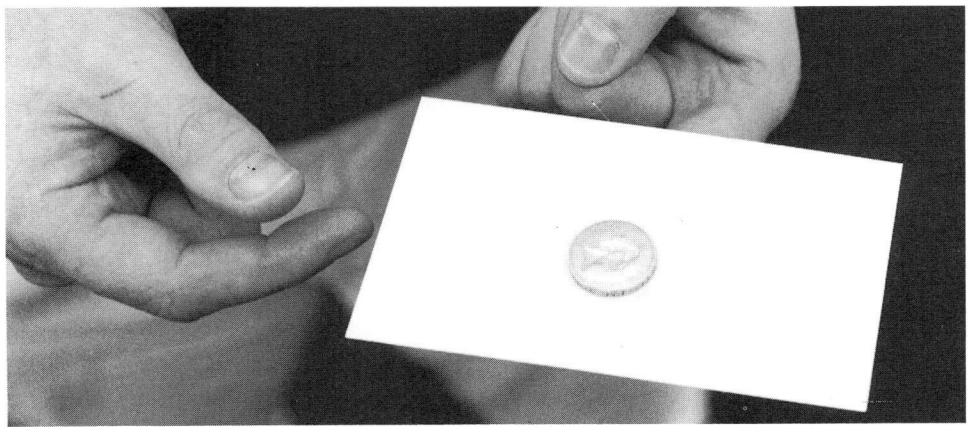

Problem 12

A ball is thrown vertically upwards and its path is shown in the diagram. The ball rises up through the point A until it reaches its highest point B. It then falls vertically downwards through the point C. Mark on the diagram, at each position, an arrow indicating the direction of the force acting on the ball. (Ignore air resistance.)

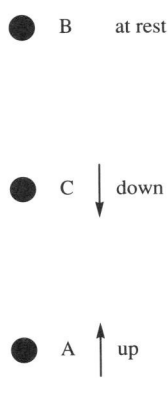

Problem 13

A ball rolls along the track and then moves through the air, following the path shown in the diagram.

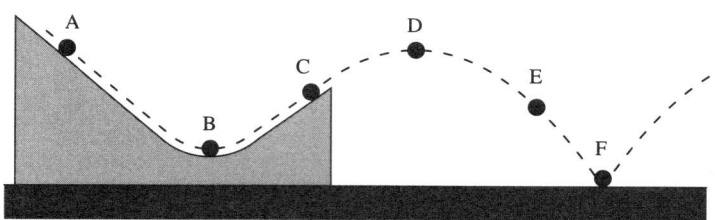

Copy the diagram and draw arrows to indicate the direction of the resultant force at each position. (The track is straight except for the lowest part where it is circular.)

Problem 14

A juggler throws two balls, one vertically and the other in a parabolic path. The dotted lines in the diagrams show the paths of each ball and the arrows show the direction of motion at points on the paths.

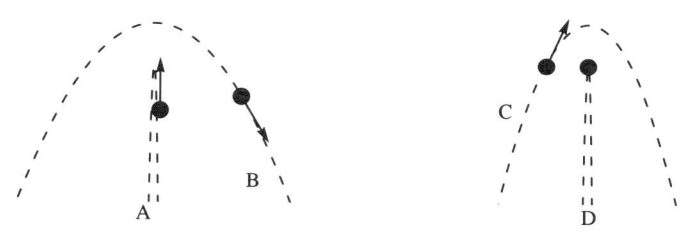

Copy the diagrams and draw arrows to indicate the direction of the resultant force on each ball at the positions shown.

Problem 15

The pendulum in the diagram below swings from left to right through the points A and B. Copy the diagram and mark on it arrows to show the forces acting on the pendulum bob at these positions. (Ignore air resistance in this problem.)
What happens if the string is cut when the pendulum bob is
(a) at B or (b) at C? (C is the highest point.)

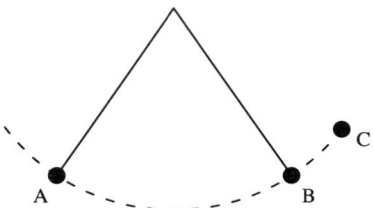

Problem 16

When a ball is thrown in the air, are the effects of gravity the same regardless of the mass of the ball?

Problem 17

Two identical objects are connected to the ends of a light inelastic string which passes over a fixed pulley as shown.
What happens if the system is released from rest?

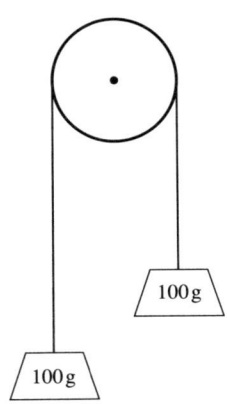

Problem 18

Two objects of different mass are connected to the ends of a light inelastic string which passes over a fixed pulley as shown.
What happens if the system is released from rest?

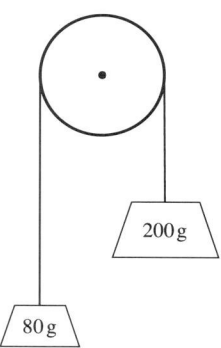

The objects are in motion when the string suddenly snaps. Describe the subsequent motion of each of the objects.
If the string snaps when the objects are on the same horizontal level, will they be on the same horizontal level again before hitting the ground?

Problem 19

Three objects are connected with strings and two pulleys as shown in the diagram. The top pulley is fixed, and two of the objects together weigh the same as the third.

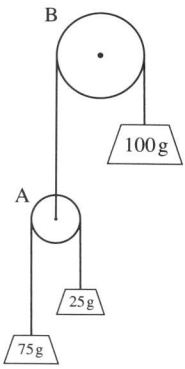

What happens if the system is released from rest?
(You should ignore the masses of the pulleys.)

Problem 20

Two objects are hanging by strings as shown in the diagram. What happens if the upper string is cut?

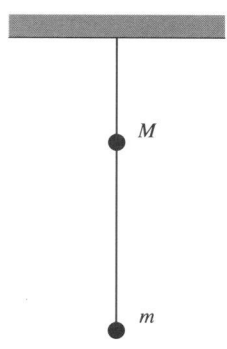

Problem 21

Suppose the strings in Problem 20 are replaced by elastic springs. What happens now if the upper spring is cut?

Problem 22

A coal wagon of mass M is hauled up a 30° incline by means of a rope passing over a pulley and attached to a tank of water of mass m which can drop down a vertical shaft.
(a) What happens if the tank of water is released? In particular, what happens if (i) $m = M$, (ii) $m = \frac{1}{2}M$, (iii) $m = \frac{1}{3}M$?
(b) What happens, in each of the three cases, if the rope snaps a short time after the system is released? (Assume that friction can be ignored in this problem.)

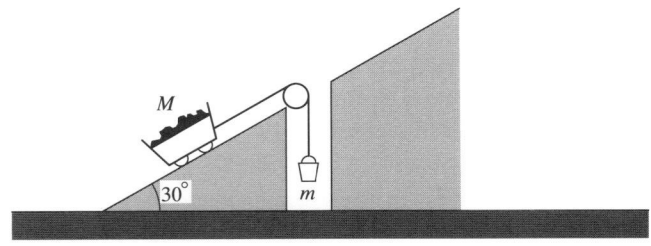

Problem 23

A solid wooden cube of side 5 cm rests on a flat table with one edge touching a length of beading attached to the width of the table.
What happens if end A of the table is gradually lifted? What happens if end B of the table is gradually lifted?

Problem 24

My friend says:
'My dog does not need dog bars to keep it in the back. It stays pinned to the back door if I go fast enough!'
Do you agree? Why?

Problem 25

A van, full of budgerigars in cages, is driven on to a weighbridge. The driver notices that the van is overweight. Fortunately, the weighbridge operator has his attention distracted and the driver realises that if she can reduce the weight she will be in the clear. She has a brainwave and says to herself 'If I sound the horn all the birds will take off and the weight will be reduced.'
Does she get away with it? Explain your answer.

Problem 26

A girl throws a ball at a wall and then runs towards the wall. By drawing possible paths of the ball, decide if it is possible for the girl to catch it.

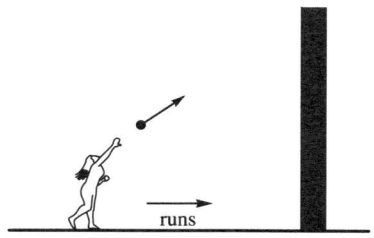

Problem 27

Each of the following objects is projected with the same initial velocity across an ice rink:
a bean bag
an ice hockey puck
a scrubbing brush (bristle side down)
a scrubbing brush (wooden side down).
Put them in ascending order of the distances you would expect them to travel before coming to rest.

Problem 28

Place an object on a table and lift one edge until the object just begins to slide. What *could* happen if the edge is lowered? Explain why the effect(s) takes place. Draw a diagram showing the forces on the object
(a) when it is at rest,
(b) when it is moving.

Problem 29

A box of mass 10 kg is placed in the back of a removal van. The coefficient of friction between the box and the floor is 0.5. What happens to the box if the lorry moves off with an acceleration of
(a) 4 m s^{-2},
(b) 6 m s^{-2}?
(Take $g = 10 \text{ m s}^{-2}$)

Problem 30

Two boys try to stretch a spring using the two methods shown in the diagram.
By which method is it easier to stretch the spring?

Problem 31

A particle is attached by springs to two fixed points A and B. The particle is held at the mid-point of AB and released. What happens when the particle is released if
(a) the springs have the same natural length and different stiffness,
(b) the springs have the same stiffness but different natural lengths?

Problem 32

An object hangs by a spring from a fixed point O. The object is pulled down from its equilibrium position A and released so that it oscillates between B and C.

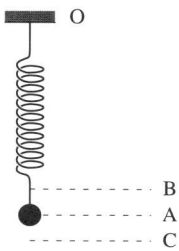

Draw three diagrams showing the resultant force on the object in each of the three positions A, B and C.
The object suddenly comes off the spring. What happens if the object comes off at
(a) point A going up,
(b) point A going down,
(c) point B,
(d) point C?

Problem 33

An object is oscillating as in Problem 32. Will the object oscillate faster or slower if
(a) the mass of the particle is doubled,
(b) the spring is replaced with another spring of the same natural length but with double the stiffness,
(c) the spring is replaced with another spring with the same stiffness but double the natural length?

Problem 34

Two identical objects each hang by two identical springs from a fixed support, as shown in the following diagrams.
Does the object in experiment 1 oscillate faster or slower than the object in experiment 2?

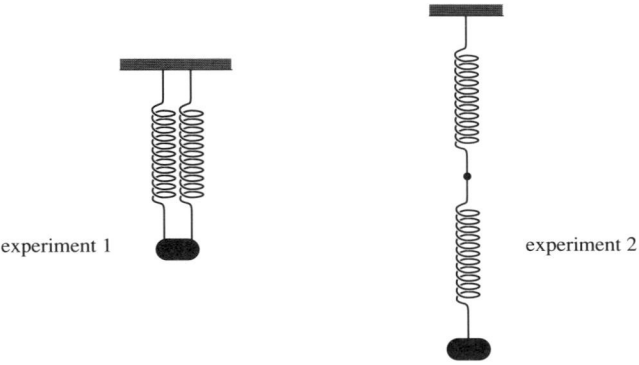

Problem 35

A small object attached to one end of a string is swung around in a vertical circle.
What will happen if the string is cut when the object is at
(a) point A,
(b) point B,
(c) point C,
(d) point D?

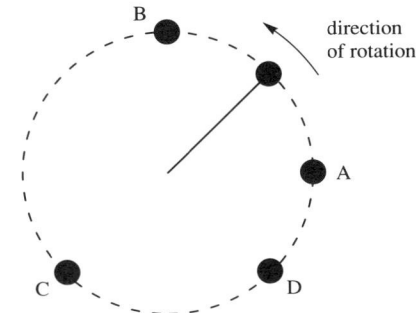

Problem 36

The object in Problem 35 is swung in a horizontal circle. What happens in this case when the string is cut?

Problem 37

Two children are sitting opposite each other on a rotating roundabout. One child aims a ball directly at the other and throws it? What happens?

Problem 38

A car is released from rest on the looped track shown in the photograph. What could happen to the car? What does the outcome depend on?

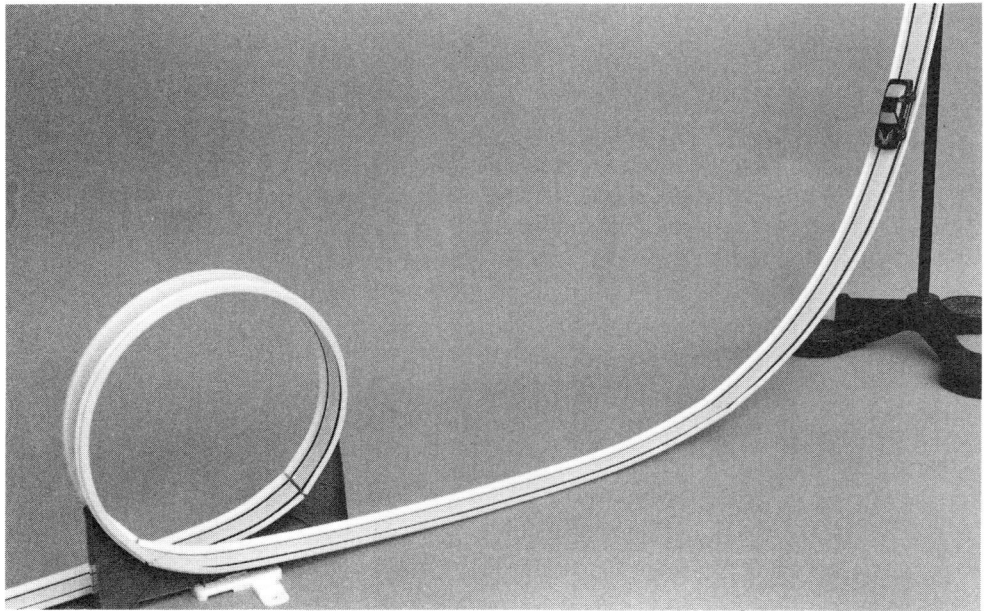

Problem 39

What happens if an astronaut lets go of a spanner
(a) while standing on the surface of the Moon,
(b) while in an orbiting spacecraft?

Problem 40

An astronaut is in a circular space station which is far from the effects of any planet and which is revolving at 1 radian per second to simulate gravity.
He lets go of a ball in front of him from a height of 2 metres. Where would the ball land in relation to the astronaut? Draw the path that the ball follows.

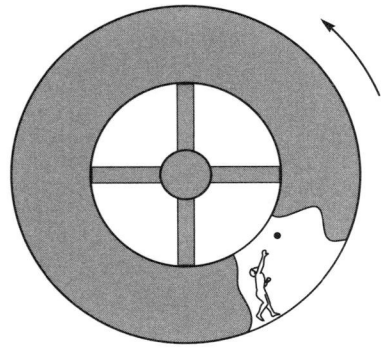

Problem 41

What happens if a penny is placed on a turntable and the speed of rotation is steadily increased? Describe the path of the penny.

Problem 42

The photograph shows a conical pendulum in which the object travels in a horizontal circle so that the string makes an angle θ with the vertical.
What happens to the angle θ if the mass of the object is increased?

Problem 43

A fairground ride involves two sets of independently rotating arms. Mark in the points where you would experience the maximum and minimum force if
(a) the main rotor is moving clockwise and the smaller one anti-clockwise,
(b) both are moving anti-clockwise.

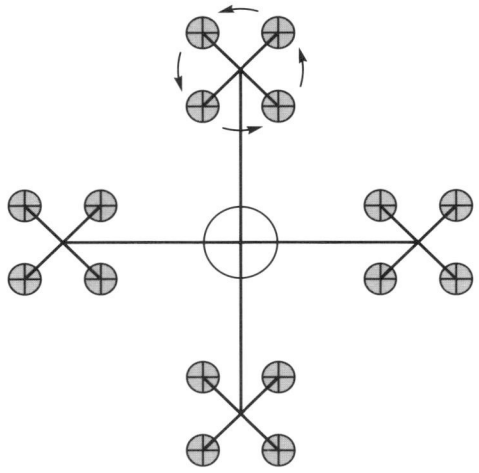

Problem 44

An easy way of making money at a school fete is the 'hit the bottle game'. The challenge offered is to swing a ball hanging on a piece of string so that it hits a bottle.
However, the game is a little harder than it sounds! To show your skill you must miss the bottle on its forward motion and hit it from behind on its return. Is it possible to hit the bottle? Explain your answer.

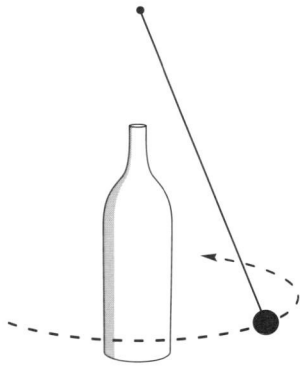

Problem 45

Which would hurt more and why?
(a) You are hit by a soft ball travelling with speed 10 m s^{-1}.
(b) You are hit by a hard ball with the same mass and size as the soft ball and also travelling with speed 10 m s^{-1}.

Problem 46

The photograph shows a device known as Newton's cradle. Each of the spheres is of the same mass and size. The suspension strings are inelastic.
(a) What happened after the photograph was taken? Describe the initial and subsequent motion of the system.
(b) The motion eventually ceases. Why?

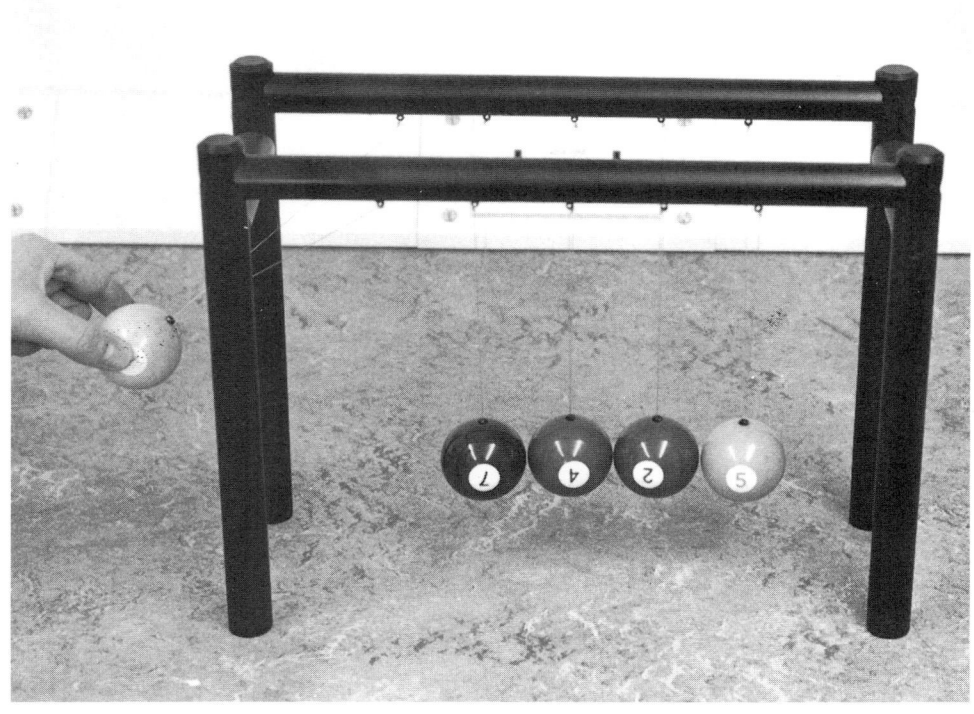

Problem 47

Perform the following experiments. Describe and then try to explain the results. You will need several 1p and 2p coins and a smooth table (preferably formica-topped). The coins will slide better 'heads' side down.

Experiment 1
Lay out five 1p coins in a straight line (line them up with a ruler) and line a sixth 1p coin up a short distance away. Flick the sixth coin towards the line of coins trying to hit the end one dead centre.

Experiment 2
Lay out the coins as above, but with a gap in the line of five. Flick the sixth coin in the same way.

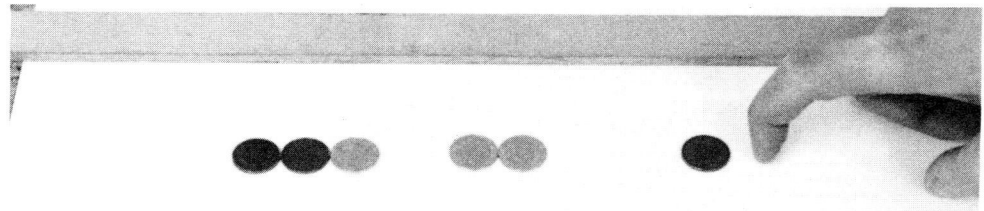

Experiment 3
Now repeat experiment 1 with a 2p coin at the head of the line.

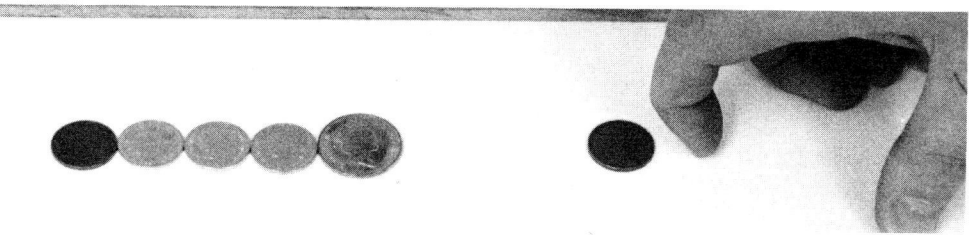

Experiment 4
This time flick the 2p coin at the line of 1p coins.

Now set up 1p and 2p coins in various combinations and investigate the collisions.

Problem 48

What happens to pool ball B if pool ball A is hit by a third ball in the direction indicated ?

Problem 49

A cue ball rolls and strikes a stationary red snooker ball. The collision can be considered as perfectly elastic. According to the laws of collisions the cue ball should stop dead; but it usually follows the red ball. Explain why.

Problem 50

What happens if a small superball (an extremely bouncy rubber ball) is dropped on top of a large superball, as shown in the diagram, so that both balls drop together?

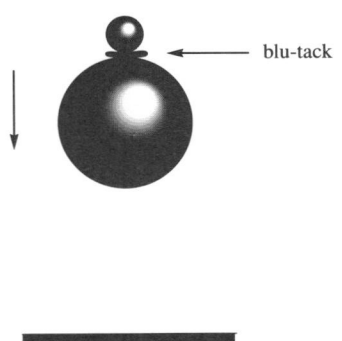

Problem 51

Two objects of mass 1 kg and 2 kg at rest on a smooth table are connected by a string which passes over a frictionless pulley. What happens if you pull the pulley in the direction shown?

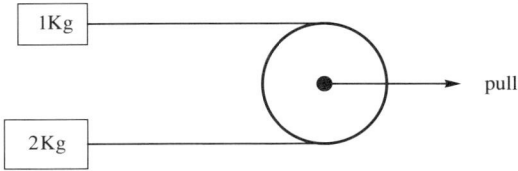

Problem 52

(a) Standard building bricks 24 cm wide and 8 cm high are stacked with each brick overhanging the one below by 8 cm. What eventually happens?

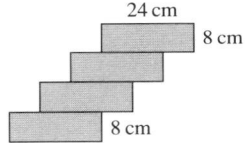

What happens if
(i) two bricks, (ii) three bricks, (iii) four bricks are stacked?

(b) The bricks are glued together instead of being stacked. What happens if
(i) two bricks, (ii) three bricks, (iii) four bricks, (iv) five bricks are glued?

Problem 53

(a) A piece of A4 card hangs from a fixed point by a string attached to the card at some given point as shown in the diagram on page 21.

How will the card rest when the given point is
(i) A, (ii) B, (iii) C?

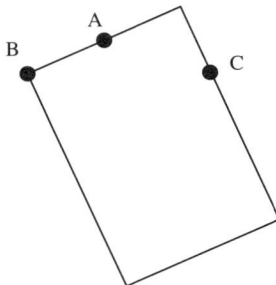 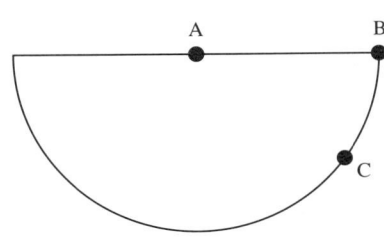

(b) What happens for the semicircular card in the diagram above when the string is attached at (i) A, (ii) B, and (iii) C?

Problem 54

A child and an adult are sitting on a kitchen bench as shown in the diagram. The bench weighs 200 N. If the child stands up, what happens if the adult weighs
(a) 400 N,
(b) 500 N?

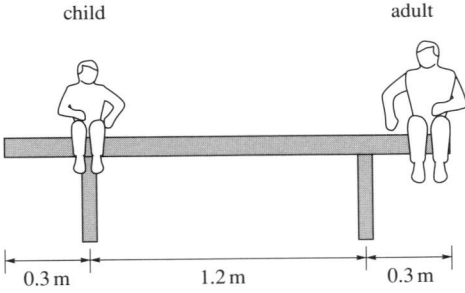

Problem 55

A step ladder stands on a highly polished floor. The two parts of the ladder are connected by a rope. A man stands on one side of the ladder, more than half way up.

What happens if the rope is cut?

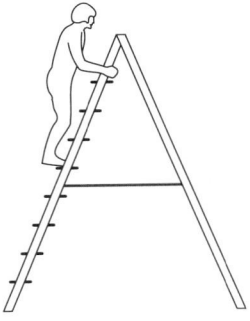

Problem 56

The diagram shows the top view of a windsurfer.
Why does it turn to the right if the sail is pushed forward?

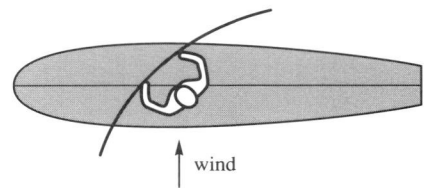

Problem 57

Helicopters usually have a horizontal main rotor and a vertical subsidiary rotor. What happens if
(a) the subsidiary rotor stops in flight,
(b) the main rotor stops in flight?
We know that it may crash but what is the immediate effect on the motion of the helicopters?

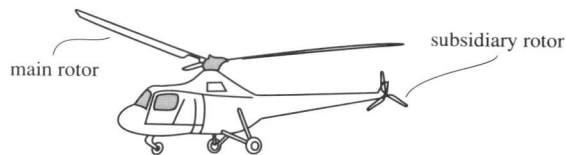

Problem 58

An ice skater revolves on a spot with her arms extended at right angles to her body. What happens if she draws her arms into her sides? Why?

Problem 59

A full drinks can and an empty drinks can are placed side by side at the top of a slope, released, and allowed to roll down. Which can, if either, will get to the bottom first?
Suppose a larger container, for example one containing custard powder, was allowed to roll down at the same time. What happens now?

Problem 60

A carton of milk stands on the floor of a car. It is observed that if the carton is full, or empty, it sometimes falls over as the car goes round a bend in the road. What happens as the level of milk in the carton varies?

PART 2

Short Investigations

Investigation 1: The two-second rule

'Only a fool breaks the two-second rule'

This catch phrase was used in an advertisement on television to encourage sensible motorway driving. In the advertisement a driver is told to pick out a feature, such as a motorway bridge or breakdown telephone, and take note when the car in front passes this fixed position.

If you cannot complete saying the phrase 'Only a fool breaks the two-second rule' before reaching the fixed position, then you are driving too close.
Why is two seconds significant?

The phrase takes about two seconds to say (try it and see), which suggests that if you are more than two seconds behind the car in front, then you should be able to stop safely in an emergency.

Is this a good rule for all speeds?

Can you suggest a better rule?

Investigation 2: An experiment for reaction times

When an event occurs, the brain takes a certain time to react to it. For example, when driving along a road, a driver can take roughly two thirds of a second to react to the car in front beginning to brake. This time is called the *reaction time*. How far would a car travelling at, for example, 50 mile/h go in this two thirds of a second? It is important for drivers to allow for the time they take to react when driving, and to be aware of the fact that they may have slower than average reaction times.

A simple experiment to measure reaction times can be carried out using a metre ruler. A friend holds the metre ruler vertically and drops it, while you try and catch it as soon as possible after it is released.

Measure the distance that it has fallen and, from this, estimate your reaction time. Construct a table from which other people can read off their reaction times. You could also use these figures to calibrate a ruler to find reaction times directly.

The Highway Code includes a table of stopping distances based on thinking distances and braking distances. Calculate the thinking time that these figures are based on, and compare it with your results.

Do you have any comments?

Investigation 3 : Motorway traffic jams

Traffic queues often form on motorways when three lanes of traffic are forced to reduce to two lanes or one lane because of road-works. This is a major frustration to motorists and often extends journey times by hours.

Consider the problem of reducing two lanes of traffic to one. What speed of traffic achieves a maximum flow through the motorway section?

The Highway Code recommends the following distances between vehicles for different speeds.

Shortest stopping distances – in metres and feet							
Speed (mile/h)	Thinking distance (m)	(ft)	Braking distance (m)	(ft)	Overall stopping distance (m)	(ft)	On a dry road, a good car with good brakes and tyres and an alert driver will stop in the distances shown. Remember these are shortest stopping distances. Stopping distances increase greatly with wet and slippery roads, poor brakes and tyres, and tired drivers.
20	6	20	6	20	12	40	
30	9	30	14	45	23	75	
40	12	40	24	80	36	120	
50	15	50	38	125	53	175	
60	18	60	55	180	73	240	
70	21	70	75	245	96	315	

Investigation 4: Road humps

A common way of discouraging motorists from driving at excessive speeds on housing estates, hospital grounds and other restricted road systems is the installation of road humps (commonly called speed bumps or sleeping policemen). We can assume that the road hump is some form of surface irregularity which is sufficiently uncomfortable to car and driver to compel the driver to slow down to pass over it. The highway engineer must decide where to place the road humps.

For different maximum speeds between road humps, how far apart should the road humps be placed?

Investigation 5 : The amber gambler

When approaching traffic lights the Highway Code makes a series of recommendations:

Amber means stop at the stop line. You may go on only if the amber

appears after you have crossed the stop line, or you are so close to it that to pull up might cause an accident.

From time to time drivers ignore this recommendation and accelerate through the lights when they are on amber.

Discuss what risks this approach might involve.

One danger is a possible collision with a vehicle crossing from the other direction.

Find a relationship between the critical speed and distance for which it is safe to cross the junction when the lights turn to amber.

Investigation 6 : Design a road-block

A road-block at a border post is to be made using strong tubular steel for the pole (mass per unit length = 5 kg per metre), two posts 5 metres apart at a height of about 1 metre, and some kind of counterweight.

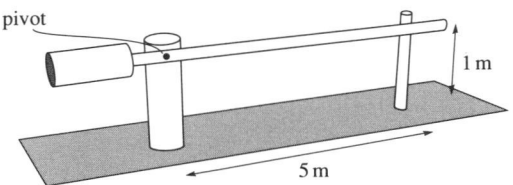

Two kinds of counterweight are available (though you may consider other kinds that you have seen or know of):

(a) A cylindrical concrete counterweight can be bolted on to the steel pole; you choose the mass.
(b) A solid steel bar can be bolted on; this has a mass per metre of 95 kg.

Your report should include diagrams indicating lengths, masses and forces. You should describe any designs considered and decisions taken.

Investigation 7: Bird-brained mechanics

An ornithologist tells me that birds fly in such a way that they always head towards their objective. A sparrow in a tree wishes to return to its nest in another tree some distance away. Unfortunately a cross-wind prevents it from flying in a straight line.

As a specific example suppose that the trees are 20 m apart, the bird can fly with a speed 6 m s^{-1} and the cross-wind has a speed 4 m s^{-1}. Determine the path followed by the bird and the time taken.

Investigation 8: Artificial gravity

It is widely accepted that orbiting space stations will need to provide an environment which has artificial gravity if personnel are to be able to stay aboard for extended periods. A long period of weightlessness leads to difficulties in balance adjustment on return and problems with muscles and bones losing calcium.

The most commonly suggested design for such stations is a rotating wheel so that the areas around the rim have an artificial gravitational force acting outwards, i.e. the outer surface is the floor, while the hub provides a working area for low gravity production and experimental work.

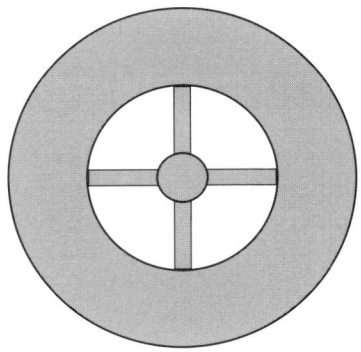

(a) If the artificial gravitational pull is to be the same as normal Earth gravity, how must the speed of rotation change if the diameter of the space station is varied?

(b) What would an astronaut experience in moving from the living area to the central, low gravity area? Why, even in the artificial Earth gravity is weight lifting easier than on Earth?

Investigation 9 : Down like a lead balloon

(An investigation into the effects of resistance.)

The following two examples illustrate the effects of resistance on falling bodies.

1 An object dropped in water falls more slowly than in air.

2 The speed of fall of a parachutist is reduced when his parachute opens

To examine the significance of resistance and establish relationships that may exist between parameters and the resistance forces perform the following experiments.

Experiment 1
Drop weights into a tank of water or a more viscous liquid such as washing-up liquid.

Experiment 2
Attach weights to a balloon and allow it to fall from some fixed points.

In both these cases measure distances fallen and times elapsed to decide whether it is reasonable to ignore the effects due to resistance.

Formulate a relationship between resistance and speed.

Investigation 10 : Bicycle gearing

A bicycle can be viewed as 'a device for enabling us to progress forward faster than when we walk, without moving our legs any faster'.

Find a relationship between the speed of the cyclist's feet and the speed of the bicycle. Your formula may contain the radii of the rear wheel, the gear wheel, the clanger and the crank. Measure these radii and find the ratio of the bicycle speed to the speed of the cyclist's feet.

PART 3
Long Investigations

Investigation 11 : The sliding ruler

Hold a metre ruler horizontally across your two index fingers and slide your fingers smoothly together, fairly slowly.
What happens?

Use the laws of mechanics to explain what you observe.

Investigation 12 : Tumble driers

In a tumble drier the clothes spend part of their time travelling on the wall of the drier and part of the time falling back through the warm air (like a projectile) towards the bottom. The design of the tumble drier should be such as to dry the clothes as quickly as possible.

One important feature of the tumble drier is the speed of rotation.
What angular speed should a tumble drier have?

Investigation 13 : Design an accelerometer

The *Drive* programme on television featured an exercise for assessing the performance of drivers. A car had an open hemispherical bowl attached to its roof in which was placed a small ball. As the car experiences different accelerations the ball will move from its position at the bottom of the bowl. Clearly the more careful driver will brake and accelerate less dramatically and the ball will as a result remain in the bowl. The driver who takes less care may force the ball to leave the bowl.

Using a hemispherical bowl and a small ball, design and calibrate a device suitable for measuring steady accelerations.

Investigation 14 : Water chute mechanics

Why is there water on a water chute?

Waterworld UK are designing a straight chute 100 m long dropping into a pool. What range of values of the angle to the horizontal are feasible so that the person slides down and hits the water at between 3 and 10 m s^{-1}?

As an experiment, use dry ice in a plastic container with a lid and with holes in the bottom to simulate water between the container and a surface. (Dry ice is readily available from some large supermarkets.)

The Kamikaze run in Spain is a water chute that consists of two straight sections at an angle; the first section is very steep and the second relatively shallow.

What is the purpose of each section?

Design one such water chute bearing in mind that the maximum safe speed of entry into the pool is 10 m s^{-1}.

Investigation 15 : Convoys

It is a well known phenomenon that when a line of vehicles is moving in a convoy (army trucks, funeral processions etc), even if the leading vehicle is driven at a fairly close approximation to a fixed speed, the latter vehicles find that most of the time they have to go significantly faster or slower in an effort to stay the same distance apart.

Investigate the effect on a convoy of vehicles of variations in the speed of the first vehicle. Can you recommend a limit to the length of the convoy?

Investigation 16 : Smashing windscreens

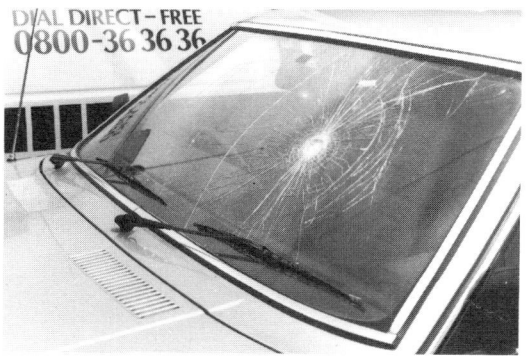

When cars travel along a road they should always stay at least $\frac{2}{3}$ second apart, even at low speeds. (This is the minimum allowance for thinking time, called the reaction time.)

However, on newly made up roads loose gravel can be thrown up by the car in front and damage the paintwork, or even shatter the windscreen. This can be very dangerous. Consider some of the following strategies to avoid this danger:

1. Set a speed limit so that the gravel always falls in front of the car behind. What should the speed limit be?

2. Recommend that drivers keep their distance, i.e. that they maintain a separation equal to the Highway Code stopping distances, given by $d = 0.682v + 0.076v^2$, where v is in metres per second and d is in metres. Is this safe at all speeds?

3. Protect all cars with a new laminated windscreen designed and tested to withstand impacts. Work out the impact speed which the windscreens should be designed to withstand.

4. Set a speed limit as in **1** but so that the gravel cannot reach the *windscreen* of the car behind.

Investigation 17 : Touchdown

The quarterback in American Football governs much of the offensive (attacking) side's play. He pre-determines where he wants his receiver to run during the play. On getting the ball from the line of scrimmage, he drops back giving himself time to survey the defence, and giving his receiver time to get in position. He is rarely able to pass the ball directly to a receiver for a touchdown; he is more likely to have to judge to which open space on the field he can throw the ball so that a receiver can get there at the same time as the ball does.

Investigate whether touchdown passes are based on judgement or pure luck.

Investigation 18 : Catch it!

In the game of cricket one of the most exciting events occurs when a player in the outfield makes a run to take a catch. Suggest a strategy with which a fielder can make sure that he arrives at the right position to take the catch.

Investigation 19 : Parachutist's dilemma

This problem concerns the motion of a parachutist from the instant of leaving the aircraft until the parachute is opened.

When the parachutist jumps from a slow flying aircraft the parachute is opened after a short free-fall, just sufficient to clear the aircraft. However, slow flying aircraft in straight and level flight are ideal targets for attack from the ground and it is therefore desirable for aircraft to fly at high speeds. Then the impact force on the parachutist when the parachute is opened can be excessive.

It is claimed that the velocity of the parachutist will pass through a minimum value less than either his initial or terminal velocities, and it is at this instant of time that the parachute should be opened.
Investigate the validity of this statement, and suggest a time when the parachute should be opened.

Related Information:

The suggested range of speeds for the aircraft are 110 to 225 m s^{-1}.

Measured values of terminal velocity indicate that it ranges from 67 to 78 m s^{-1}, approximately.

Investigation 20 : Sprint starts

There are several different sprint start techniques available using starting blocks (see AAA Coaching Manuals). They are classified according to the toe-to-toe distance. In increasing order of this distance they are the bullet, medium and elongated starts.

You have been approached by a sprinter and asked for advice as to which is the best technique to use. Describe how you would set about investigating this problem, the data you would collect and how you would use it to produce your final advice for the sprinter.

Note: Although the toe-to-toe distance is obviously dependent upon the stature of the athlete the following figures are representative for a senior male athlete:

bullet   28 cm
medium   40–54 cm
elongated 60–72 cm